Not Just Night!

Nocturnal Helpers

Written by Diane Kann

Illustrated by JB Barbara

Paws to Read – A moment shared is a memory made.

Visit us at: ***boltandcharliebooks.com***

Published by **Kannceptual Creations LLC**

An imprint of Bolt and Charlie Books

boltandcharliebooks.com

ISBN: 978-1-972343-51-7

Printed in the United States of America

First Edition, April 2026

Paws to Read is a Bolt and Charlie Books initiative promoting early literacy and shared storytime experiences.

Dedication

To all children, near and far,
who sleep beneath the same bright stars—
this book is for you,
to remind you that even in darkness,
there is light, life, and kindness.

When the sun sinks low from sight,
who helps the world throughout the night?

You might think all's quiet, still...
that everything sleeps beyond the hill.

But wait! Step softly—take a peek...
nighttime helpers start their streak!

An owl glides on silent wing,
listening close for every thing.

It helps keep balance in the land,
a nighttime hunter, strong and grand.

A bat swoops low through moonlit air, snatching insects here and there.

It helps control the bugs you see,
a fluttering friend to you and me!

A moth drifts softly, pale and light,
visiting flowers in the night.

From bloom to bloom it gently goes,
helping nighttime flowers grow!

**A raccoon searches near the stream,
washing snacks in silver gleam.**

It cleans and scatters seeds along, helping forests grow strong!

A firefly flickers—flash, flash, glow!
lighting pathways down below.

**Calling friends with shining light,
a tiny lantern in the night!**

Even crickets chirp and play,
filling night with songs that stay.

**Each small sound and glowing spark,
plays a part within the dark!**

So when you sleep and dream just right,
remember the helpers of the night...

THE END...

Goodnight Message

Sleep well, little one...
The night is alive, the helpers are near,
working with care while you dream without fear.

NIGHT FRIENDS
PLAYGROUND
- EXPLORE! PLAY! SHINE!

www.ingramcontent.com/pod-product-compliance
Lightning Source LLC
Chambersburg PA
CBHW041323210926
51847CB00002B/34
* 9 7 8 1 9 7 2 3 4 3 5 1 7 *